Pig Keeping on a Small Scale

This book is dedicated to all those enthusiasts who work towards saving endangered breeds of domesticated animals.

Pig Keeping on a Small Scale

Annette and Grant McFarlane

Kangaroo Press

Acknowledgments

Special thanks to John and Libby Mulholland for introducing us to the joys of keeping Tamworth pigs; Pat Crowther, Judy and Brian Thompson, Frances Lang and Jeff Michaels for their help and encouragement; Anne Mabey and Paul Udy for their constructive and helpful reviewing of the draft text.

LINCOLNSHIRE COUNTY COUNCIL

636.4

© J G and A M McFarlane 1996

First published in 1996 by Kangaroo Press Pty Ltd.
3 Whitehall Road Kenthurst NSW 2156 Australia.
PO Box 6125 Dural Delivery Centre NSW 2158.

All rights reserved. No part of this publication may be reproduced, stored in a retrieval system, or transmitted in any form or by any means, electronic, mechanical, photocopying, recording or otherwise, without the prior permission of the authors.

ISBN 0 86417 790 9

Printed in Hong Kong through Colorcraft Ltd

Contents

Preface	6
1 Background	7
2 Introduction	8
3 Why Pigs?	9
4 Starting Out	12
5 Pedigree Pigs	15
6 Purchasing Options	19
7 Production Systems	22
8 Feeding	24
9 Watering	35
10 Housing	37
11 Fencing	40
12 Breeding	41
13 Growth and Development	49
14 Handling and Management	51
15 Animal Health	52
16 Composting Manures	59
Glossary	60
Bibliography	61
Other Information	62
Botanical Names of Lesser-known Plants Mentioned	63
Index	64

Preface

In writing this book we hope to share our experiences in keeping pigs and encourage others to see pigs as productive, manageable and rewarding animals for individuals with small landholdings.

Newly introduced species such as alpaca, lama and ostrich are an expensive and risky investment for the inexperienced person. Pigs, on the other hand, are relatively inexpensive, easy to care for and largely free of pest and disease problems. Unlike other species whose market potential is often overestimated or unknown, pig products are easily disposed of through existing markets or small local networks.

We keep Tamworth pigs, an English breed now classified by the Rare Breeds Survival Trust as critically endangered. Enthusiasts are needed to maintain these and many other rare breeds of pigs which form the basis of today's hybrids.

These pure strains have fallen from favour over the last three decades because they have proved unsuitable for use in modern intensive production systems.

In April 1995, four Tamworth pigs from Myola Tamworth Pig and English Leicester Sheep Stud in Queensland were exported to England to provide new blood lines to strengthen genetically weakened British Tamworth breeding stock. It was the first successful export of Tamworth pigs to England in almost twenty years.

We were fortunate to be able to assist in supplying quarantine facilities for the animals and play our part in helping to save an endangered breed.

Annette and Grant McFarlane
Macaranga Tamworth Stud
Samsonvale QLD

1. Background

We are often asked, 'How on earth did you get interested in pigs?' We have our good friends and avid pig keepers, John and Libby Mulholland to thank. Several years ago John and Libby were looking for someone to house sit for them while they headed overseas for a few months. The opportunity to live in a beautiful, old Queensland home on three hectares of land in a leafy, outer Brisbane suburb was hard to resist. It was to be like a holiday in a rural homestead complete with a dog, cat, pigs and sheep.

Neither of us had any real experience with animals. We had dogs and cats as children, but I never really had to look after them. Grant knew a little about sheep as a result of childhood holiday expeditions to farms belonging to distant New Zealand relatives. Other than that we were equipped only with plenty of enthusiasm and a comprehensive list of who got fed what and how much. A list of telephone numbers attached to the refrigerator detailed who to call in case of some dire emergency.

The cat died and the dog ran away for days at a time. The sheep almost caught pneumonia when it rained for several weeks just after they were shorn. By comparison, the pigs were a breeze. We learnt to recognise and call each one by name. We collected treats of squashy grapes from the local green grocer for them, and they were bathed and brushed regularly. In other words, we were hooked.

It was not long before we had bought our own piece of rural paradise, complete with a Tamworth sow. She was 'in pig' at the time. Her name was Lisa and many litters later, she remains the aging, dominant matriarch of our pig family.

2. Introduction

Many people on small acreage fall into the trap of having animals that do not pay their way. Once you start planning your piece of paradise and express a passing interest in keeping the odd sheep or cow, you tend to become a magnet for a menagerie of unwanted species. Unwanted pet goats, sheep, geese, ducks, chickens, guinea pigs, donkeys, dogs and cats seem to be thrust upon you from all quarters. Very often small landholders make the mistake of accepting these orphans without first considering the cost associated with their upkeep.

We made a conscious decision only to keep animals that could at least pay for themselves in some way. Payment is not necessarily measured in a monetary sense, but by the contribution the animals make to the property or lifestyle of their owners. This is a useful way to examine any animal you are thinking of keeping.

The animals you choose will be influenced by factors such as climate, topography and production method. You should also investigate the intrinsic behaviour of the animal. Work out if you can supply its needs in an efficient and cost-effective manner. Work out ways to capitalise on any yields it may produce and put any natural activities it engages in to good use.

Examine Breed	Investigate Behaviours	Supply Needs	Utilise Yields/Activity	
Type	Gregarious	Shelter	Meat	Heat
Size	Social	Food	Skin	CO_2
Colour	Hierarchical	Water	Lard	Foraging/
Temperament	Omnivorous	Air	Manure	Digging
Productivity	Inquisitive	Exercise	Offspring	
Climate Tolerance		Parasite Control	Methane	

3. Why Pigs?

Pigs are incredibly productive units when compared with other species. Their 114 to 116 day gestation period and 21-day ovulation cycle enable them to produce up to five litters every two year. As an average of eight piglets per litter is not unusual, one pure-breed sow has the potential to produce forty offspring within two years. This reproductive rate may be even higher with cross-bred or hybrid pigs. The productivity of a pig is phenomenal when compared with the single offspring of a sheep or cow. This fertility has long been recognised in third world countries, where pigs provide a valuable food source. Pigs produce more meat world-wide than any other animal. They are particularly popular in China and the Pacific region.

Pigs are also more efficient at converting food into meat than any other animal. The feed to meat conversion rate for commercial pig breeds is around 3:1. In other words, three kilograms of dry feed produces one kilogram of meat. By comparison, the conversion rate for cattle and sheep is approximately 8:1.

While it is a fallacy to say that pigs will eat anything, it is possible to recycle many waste products via pigs. Kitchen scraps (excluding meat), vegetable tops and unsaleable fruit and vegetables from the local greengrocer, or stale bread from a local bakery, are all good food sources for pigs. It is also possible to plant your garden so as to be largely self-sufficient in terms of food supply for both your family and your pigs.

Pigs are also manageable for the small landholder. They can be accommodated on a relatively small area of land. Transporting large animals to market involves considerable expense. Several pigs can easily be carried on the back of a utility fitted with a steel cage or on a modified box trailer.

People who view pigs as dirty animals have probably never seen a pig or associate the habits of all pigs with the feral animals that run wild in much of outback Australia and New Zealand. Properly housed and fed, domestic pigs are no dirtier than any other animal. In fact, they are very clean. Given the space, pigs will not defecate where they eat, drink or sleep. They will defecate in one far corner of the pen. Given a good diet, pigs smell no worse than other animals. We will take pig manure in preference to dirty nappies any day.

Pigs can be kept as free-ranging animals. We keep our pigs confined to large, comfortable pens with daily access to pasture grazing. Regardless of which method you choose, it is advantageous to be able to supplement your pigs' diet with food from orchards and vegetable gardens. Pigs are useful as land clearers and will dig out blackberries and lantana in search of starchy roots.

Every gardener knows that composted manure makes an incredible difference to the growth rate of plants. Manure provides valuable organic material, improves soil fertility and is a food source for many soil organisms. Composted animal manures improve the growth and productivity of plants.

People on small acreage often plant trees, so they may as well be productive, fruiting species. Keeping pigs may be a way of using these on-site resources as they can consume any excess fruit and vegetables. Pig manure is a valuable source of organic material for fruiting trees, vines and vegetable gardens. This closed producer/consumer loop uses waste and avoids pollution.

Other Characteristics

Pigs are social animals. They enjoy the company of other pigs and get on quite well with most other animals. A solitary pig is a sad and sorry sight. Few animals worry pigs. They will graze happily with sheep, cows and goats. We have been told horses do not like pigs, but have no first-hand experience of this. Any animal deciding to take on a pig would generally come off second best. Our kelpie plays happily with young pigs, but has a healthy respect for large sows and our boar.

In subtropical and tropical areas of Australia, carpet snakes are a real problem for owners of small livestock. It is not unusual for a mature carpet snake to eat four or more chickens in one night. We know of instances where ducks and young goats have been consumed whole. Small piglets would seem an obvious target, but even carpet snakes are unlikely to attack piglets protected by a sow.

Pigs and chickens are a good combination. In our experience, pigs will not harm even the smallest chicks, despite the fact that they steal food morsels from their feed trough. Chickens will clean up anything left by the pigs and spare eggs are not only relished by pigs, but are a valuable source of protein. If kept with milk-producing animals such as cows or goats, pigs can use surplus milk or milk byproducts.

Pigs have an acute sense of smell, but have poor night vision. We once let our sow out to roam in a distant paddock, expecting her to make her way home by dark. Night had fallen when we went to feed her, but she was nowhere to be found. We could hear movement at the bottom of the paddock and had to lead her by torch light back to her pen. She could not find her way over the unfamiliar ground alone in the dark.

Children are fascinated by our pigs and always want to feed and touch them. Whenever possible we always try to keep a few pumpkins on hand for demonstration purposes. When a pig crushes a whole butternut pumpkin as if it were a grape, we point out the dangers of placing juicy, sweet-tasting fingers through the fence.

Believe it or not, pigs and humans have a lot in common. Their body temperature is similar to ours and tissues from pigs were used in human medical procedures before synthetic substitutes were developed (for example, insulin). Transgenic pigs implanted with human DNA have even been suggested as a source of human transplant organs.

4. Starting Out

There are a few things to consider before you buy your first pig. Many councils and shires do not allow pigs to be kept within their boundaries or place restrictions on the number of animals that landowners can keep. It is worth making an anonymous call to your local authority to gauge their reaction.

Some local authorities that do not allow pigs to be kept, will only act if the animals are a cause of complaint from neighbours. If pigs are out of sight and out of earshot, neighbours are unlikely to complain. If they are pork lovers, the thought of quality, locally supplied meat is often enough to silence any complaints. The odd free kilogram of bacon is always appreciated.

When people see our clean, happy, healthy pigs, they generally fall in love with them. However, people often have ill-informed, preconceived ideas about pigs. Consider your neighbours when deciding where to place pens. Screening trees and shrubs can alleviate any perceived visual problems and good management practices will help to reduce any undesirable odour or noise.

How do you intend to keep your pigs? Not all land will be suitable for free-ranging pigs. Grazing land for pigs should be free draining. It should not be too stony. Pigs are likely to suffer from foot injuries on very rocky ground. Land that is very steeply sloping is also unsuitable. While some slope is desirable for good drainage, as a guide paddocks should have a gradient no greater than 1:8.

Decide whether you are going to keep pigs primarily for meat production or whether you intend to breed and sell live pigs of particular blood lines. Remember that even if you decide only to keep pure breed pigs and sell live breeding stock, some inferior off-spring will need to be culled. You may decide to sell piglets for others to grow on, or choose to use them yourself. Either way, at some stage, inferior or ageing pigs will need to be culled.

How will you slaughter your pigs? Home killing and curing of pork and bacon should only be carried out by experienced people. Pigs must be killed quickly and humanely. Do you have the skills, the right equipment and area to hang the carcass? How will the remaining stock react? The manner in which pigs are slaughtered, chilled and prepared will significantly affect the quality of the pork or bacon produced. For these reasons we send our pigs to a small, local abattoir where we know the job will be done quickly, humanely and efficiently.

Not all butchers have an abattoir licence. You may have to travel some distance to find someone who is able to do the job for you. In some areas, mobile butchers are available who will come to your property. It is best to ascertain this before buying your stock.

A little less than one-third of the weight of a live animal is lost in processing the meat. For example, a porker with a live weight of 60 kilograms will produce approximately 43 kilograms of dressed pork. A baconer with a live weight of 100 kilograms will produce approximately 75 kilograms of meat. This will comprise eight large hams (it is usual to cut each leg into two) and the remaining weight in bacon.

We have found the cost associated with having animals professionally butchered is quite reasonable. We currently pay around $1.00 per kilo for pork and $1.50 per kilo for sugar-cured ham and bacon (dead weight).

Pork can be prepared in traditional cuts like roasts and chops, or the so-called 'new-fashioned pork'. The latter requires porkers to be grown on to a much larger size — between 80 and 95 kilograms.

The meat is returned packaged ready for freezing and boxed in single sides. The butcher is also able to give us feedback on the quality of the meat. This is particularly important when you first start out or are experimenting with different food sources. The butcher will be able to advise on the size and quality of the animals supplied and whether the meat is sufficiently lean.

How will you dispose of the excess production? One sow, regularly mated, will produce more pork, bacon and ham than any one family could possibly eat. If you develop a reputation for good quality, lean animals your butcher may be prepared to buy from you. Alternatively, consider whether you have a network of friends or acquaintances who would be interested in buying or trading goods and services for meat. This is our major avenue for disposing of excess production.

Pork prices fluctuate according to the season and supply and demand. Organic pork is almost unobtainable in Australia due to the limited sources of organic grain and requirement that abattoirs conform to stringent certification guidelines in relation to use of chemical disinfectants. Some consumers are prepared to pay premium prices for organic meat produced by small local enterprises. However, others think that the pork should be cheaper than at the butcher because your overheads are lower. In setting a price for your pork, you must take into account all the costs associated with keeping the animals including housing, feeding, transport, abattoir and butchering costs.

Many people choose not to eat meat because of the hormones, growth enhancers, antibiotics and other additives it may contain. Of particular concern to pork consumers in Australia has been a recent ruling by the National Food Authority that pork from pigs treated with the controversial growth hormone, porcine somatotropin, can be marketed without any special labels to advise that the meat has been treated in this way. While the use of porcine somatotropin has been rejected by the majority of Australian pork producers, its use still remains an option. Porcine somatotropin is not used anywhere else in the world.

Other consumers view some aspects of intensive commercial animal production as inhumane. We have never had any difficulty disposing of excess pork to these and other consumers. They will quite happily buy our pork, bacon and ham knowing that it is free from artificial contaminants and antibiotics and that the animals have been housed and treated humanely.

Ham and pork are always popular during Easter and at Christmas. With a little planning it is possible to time the development of litters for such events. Details of how this can be achieved are given in the section on planning production (page 41). People often comment on how different our pork tastes. Many have remarked on its flavour, noting that it takes like 'real' meat used to.

Most freezer manufacturers only recommend storing meat for one to three months. However, we have successfully stored pork, ham and bacon for at least twelve months without noticing any deterioration of the product. Salt tends to crystallise on the outside of frozen ham and bacon. If this is excessive, simply wash the salt from the meat using clean water. Our 150 cubic litre chest freezer will hold four porkers.

If you decide to keep pigs you will need to provide them with daily attention, water, housing and fencing. None of this is designed to discourage you from keeping pigs. However, the pleasure of keeping animals comes with a certain degree of responsibility. If you cannot provide their needs and use their outputs, you should reconsider keeping pigs or any other animal.

5. Pedigree Pigs

It will surprise some people to discover that just like dogs, cats, cattle and other animals, it is possible to have a pedigree pig. This simply means that a particular pig is a pure strain or breed whose parentage has been recorded and registered. Pedigree or pure breed lines are bred so as to maintain particular breed-specific characteristics and behaviours.

The pedigree name applied to a pig is usally a combination of the name of the property where the animal was bred, the name of the sow or boar line and a registration number.

Few pure-breed pigs are suitable for modern intensive production methods, hence the trend to modern hybrid pigs that can be kept entirely indoors. Hybrid pigs have been selectively bred to produce the maximum growth in the shortest possible time. Some breeds of pigs have now become extinct. Many others are critically endangered, relying on a few isolated breeders to keep the remaining pure bloodlines going.

The type of pig you decide to keep will be depend largely on what you intend to use the offspring for, what breeds are available to you, personal preference, and the conditions under which the animal will be kept. Not all pig breeds are suitable for free ranging. Light-coloured pigs such as large whites tend to get sunburnt easily.

Advantages

If you are going to keep a pig it may as well be something special and interesting. Should you choose to keep a rare breed you will be helping to preserve an animal in danger of extinction because it does not conform to the requirements of intensive, commercial pig production.

Quality stock of any species is always in demand. The rarity of certain pig breeds ensures that there will always be a market for selected breeding lines. The absence of ailments such as foot and mouth disease in Australia means that quality breeding stock can be more easily exported.

Rare breeds of pigs often have a hardy constitution. They are more resistant to disease, are less sensitive to environmental change and generally more suitable for free ranging than hybrid pigs bred specifically for intensive commercial pig production.

Disadvantages

Stock of rare breeds or pure bloodlines of pigs will be more expensive to purchase. It also takes experience to identify and select animals with the best characteristics of the particular breed. You will need to monitor and record any breeding of stock that you carry out. You will also need to join an association if you wish to register your pigs as pedigree stock.

Pure breeds of pigs may produce fewer offspring and take longer to reach predetermined weight targets than hybrid pigs bred specifically for those characteristics.

Pig Breeds

A number of pig breeds have become extinct since the introduction of intensive commercial pig production and as a result of changing consumer demand. Detailed information on lesser known breeds can be found in Porter (1987).

English Breeds

Berkshire

This black pig has a white-tipped tail and white feet. Particularly hardy in cold climates, they are good mothers and quick maturing. Unlike other black pigs, the Berkshire does not have the dark discolouration to the meat which is unattractive to some consumers.

British Saddleback (Wessex & Essex Saddleback)

This black pig has a distinctive white belt around the body. It is suitable for pork or bacon although the meat tends to be slightly discoloured. It is often crossed with Large Whites to overcome this problem. Lop-eared, hardy, good grazers and foragers, they are suitable for cold climates.

Gloucester Old Spot

The Gloucester Old Spot has floppy ears and black spots over a white body. They are well adapted to cold climates, prolific breeders and good mothers.

Large Black

The Large Black is hardy in cold climates and a prolific breeder with good mothering instincts. It does well on pasture and is often crossed with other breeds for pork and bacon production.

Tamworth

Closely related to wild hogs, the Tamworth is one of the oldest known breeds of pigs. They are hardy pigs and their golden red colouring makes them an ideal outdoor pig. They possess good temperament, mothering and breeding characteristics.

Large White

Large, long-sided, white pigs that are often crossed for use in factory farming.

The breed is noted for large litter production. It is unsuitable for continuous outdoor grazing in hot climates unless provided with adequate shade as it is prone to sunburn.

Middle White

This hardy, dual-purpose pig is suitable for pasture grazing in mild climates or confinement in housing.

American Breeds

Hampshire

Hardy black breed similar to the British Saddleback, but the white belt is narrower. Solidly built, docile and suited to both outdoor and indoor production systems.

Duroc

Reddish brown pig originally bred for lard, but now developed to be suitable for bacon. Docile and hardy under a range of conditions, they are popular for cross-breeding.

Other Breeds

Landrace

This lop-eared European breed is well suited to intensive indoor production. It is a lean, white, long-bodied pig.

Micro-pigs

West Africa, South Asia, the East Indies, Latin America and islands within the Pacific Ocean are home to some of the smallest pigs in the world. These micro-pigs generally have a mature weight of less than 70 kilograms and are renowned for their prolific rates of reproduction and hardy constitution. These pigs are often an integral component of third world subsistence agriculture, where they thrive on the byproducts of primary production and provide villagers with a valuable supply of meat.

Kunekune

New Zealand has its own unique breed of pig known as the kunekune. These small, docile grazing pigs can survive solely on a diet of well balanced pasture. They are not wild pigs but have always been kept as domestic stock. Kunekunes have a maximum weight of around 50 kilograms and grow to about 60 cm in height. An interesting feature of the breed is the dewlaps that develop on the lower jaw.

While the origins of the kunekune are uncertain they are thought to have descended from Indonesian, Asian or Chinese stock and have a long historical association with Maori people.

Once endangered, Kunekunes have recently regained some popularity with hobby farmers as orchard grazing pigs.

18 PIG KEEPING ON A SMALL SCALE

```
Name:    WOOROGLEN RANGER A8 A258204              Breed:  TAMWORTH
         (FHR A8)
Bred By:   F H ROBERTS              V 210900      Sex:   MALE
Colour:                                           Date of Birth:  30 OCT 93

                                  ┌ S. AROORA RANGER 493  26316
                                  │    (493)
                  ┌ S. AROORA RANGER V67 V250194
                  │    (V67)
                  │               └ D. AROORA BERTHA 1430 Z14229
                  │                    (1430)
       Sire: AROORA RANGER R7 R258198
            (1BA R7)
                  │               ┌ S. AROORA GLEN 543 Z32824
                  │                    (543)
                  └ D. AROORA GLEME 170 X163830
                       (X170)
                                  └ D. AROORA GLEME 1342 26333
                                       (1342)
                                  ┌ S. LESDALE ATOMIC 39 26069
                                  │    (T 920)
                  ┌ S. PARILLA ATOMIC 12 26167
                  │    (T 12)
                  │               └ D. LESDALE GOLDEN MARTHA 97 25845
                  │                    (T 695)
       Dam: WOOROGLEN GLEME 1 258202
            (FOR 1)
                  │               ┌ S. LESDALE ATOMIC 39 26069
                  │                    (T 920)
                  └ D. ROSEWORTHY GLEME 30 26027
                       (T 30)
                                  └ D. LESDALE GOLDEN MARTHA 97 25845
                                       (T 695)

                                           Issued: 19 SEP 95
                                               RPT
Registered Owner: MR G MC FARLANE          REGISTERED 18 APR 95
                  PO BOX 235
        Q 211077 FERNY HILLS   QLD 4055   TRANSFERRED ON 28 APR 95
                  Issued by: AUST. PIG BREEDERS ASSOC. LTD
                             P O BOX 189 KIAMA NSW 2533
```

Above: *Certificate of registration for a pure-breed Tamworth boar.*
Below: *Certificate of transfer for a pure-breed Tamworth boar.*

                AUSTRALIAN PIG BREEDERS' ASSOCIATION LIMITED
                P.O. BOX 189, KIAMA. N.S.W. 2533—TELEPHONE (042) 323333. FAX (042) 323350

TRANSFER [This certificate to be lodged with the Federal Secretary]

I hereby certify that the (breed) **TAMWORTH**
Boar **WOOROGLEN RANGER A8**   Herd Book No. **A258204**
~~Sow~~
Tattoo Brand **FOR**   Tattoo or ~~Ear Notch~~ No. **A8**
Born _____ Dam **WOOROGLEN GLEME 1** Herd Book No. **258202**
was sold by me to Mr. **& MRS J.G. MACFARLANE**
of **MACARANGA STUD DAYBORO QUEENSLAND** on **28th April** 19**95**
and I hereby authorise the transfer of Ownership as above to be recorded in the books of the Association.

Where this Transfer Certificate refers to a Sow, the following particulars must be supplied:
This Sow was served by the Boar **N.A.**   No. _____
on _____ 19__

Vendor's Signature *John Mulholland*
Address **Myola Stud, Pullenvale   Qld.** P/code **4069**
Vendors M/ship No. **Q210641**   **29/4/95** Date

6. Purchasing Options

The cheapest way to purchase pigs is at the weaner stage of six to eight weeks of age. Boars can be reared for a further ten weeks (sixteen to eighteen weeks in total) by which time they will have reached porker weight. Sows can be reared for an additional eighteen weeks (twenty-four to twenty-six weeks in total) by which time they will have reached baconer size. We generally do not use our boars for ham and bacon production. In non-commercial production, young boars raised with sows may become sexually mature before they have reached the required weight. This can leave the meat with a strong taste.

Buying pigs at a young age means you can train them and get them accustomed to being handled. You have more time to learn as you go. You can expose your piglets to a variety of foods from an early age. As a result they are more likely to accept a wider variety than a pig that has only been fed pellets or meal since weaning.

Pigs reach reproductive age anywhere between five and seven months, but are not usually mated until around eight to nine months. If you purchase weaner pigs and intend to rear them for breeding purposes, you will need to take into account any costs associated with maintaining them from weaner stage through to sexual maturity.

Another option is to buy a pregnant or 'in pig' sow. Although the initial cost will be greater, your stock will have multiplied considerably in just a few months when the sow gives birth. A sow in pig with her first litter is known as a maiden sow.

If you intend only to use your pigs for meat production you may be able to purchase stock being culled by breeders. These may be gilts nearing sexual maturity that do not quite match up to the breeder's expectations. A gilt is a sow that has not been mated. A variation in skin pigmentation or misshapen ear will be enough for a breeder to reject a gilt for breeding. However, these physical imperfections would not affect the ability of the animal to produce offspring suitable for meat production.

Breeders only have need of and facilities for a limited number of boars. As a result young boars are usually fairly readily available.

Breeders may also be willing to part with older sows whose litter numbers are falling. Meat from older sows can only be used for sausages, meaning that they are of little commercial value. While litter numbers are critical for the commercial breeder, they are less significant for the smaller producer. Purchasing an older sow may be a way of obtaining stock with good characteristics for a small initial outlay.

Quality Stock

Regardless of whether you intend to breed pure bloodlines or simply breed pigs for meat, it is worth selecting stock of good quality.

Poor quality stock takes just as much time to feed and care for as good quality animals, perhaps more. Pigs with poor feet and joints will be less active and graze less efficiently. This may result in slower growth rates or mean that lower rates of activity cause the animal to produce more fat than meat. Healthy, happy pigs are a pleasure to keep and require minimal care.

When selecting a pig for the first time, try to take along someone who knows what to look for. Even if the person only has experience with other species of large production animals, they will have some idea of what to look for in terms of good animal conformation.

Look for the following characteristics when selecting your pigs:

- Demonstrated good growth rate
- Medium to large sized head with broad eye and ear placement
- Clear, bright eyes
- Long straight, slightly arched back
- Deep sides and wide chest
- Strong leg and foot conformation
- Twelve to fourteen evenly spaced, well formed teats on both males and females
- Good temperament.

Where to Look and Buy Pigs

Some large agricultural shows still exhibit rare breed pigs. Prize-winning stock can often be bought after the show. Experienced judges will have sorted out the best breed lines. This may help you identify those breeders who consistently produce good quality stock. It may be possible to follow up and buy offspring of prize-winning parentage.

Rare breeds associations exist in some states of Australia and countries overseas. Contact the rare breeds association in your area for the names of pig breeders who carry the pig breed or bloodlines you are looking for.

Registered breeders take pride in their reputation for producing good quality stock. Reputable breeders will not risk selling inferior stock knowing that it will be used for breeding purposes. Only the best stock is generally retained for breeding, while pigs with inferior characteristics are sold for meat.

Obtaining Stock for Teaching Purposes

Schools and agricultural colleges offering animal husbandry programs often find that purchasing an older sow is the most economical way to obtain stock. Where possible, they arrange with a local commercial producer to collect an 'in pig' sow that is two-thirds through gestation. The students can learn about pig husbandry and watch the changes taking place in the sow in the five to six weeks prior to farrowing. Farrowing is an exciting time and following the birth of piglets, student exercises may involve monitoring growth and development.

Pigs are intelligent animals and it may be possible to undertake studies in learned behaviour. Once the piglets have been weaned the sow may be returned to the breeder or culled. The offspring could be raised to porker weight following which sale can be arranged through a local abattoir or butcher. The return on the sale of the porkers also makes it possible for the exercise to be undertaken on a cost-neutral basis.

Some commercial producers are also often willing to provide an 'in pig' sow on loan for educational purposes and provide all the food required to maintain the sow and raise the offspring. Once the students have acquired the necessary skills in pig husbandry, the animals can be returned to their owner.

With a little planning it is possible to arrange these activities to coincide with a school semester, overcoming the problems of caring for the animals during semester breaks.

7. Production Systems

Intensive Commercial Production

Pigs in intensive commercial production units are housed in pens with barely enough room to turn around. Fortunately, the tethering or chaining of sows to restrict their movement has now been banned in Australia. Within two days of birth, piglets have their teeth clipped and tails docked.

Sow milk is low in iron. This is not a problem for piglets that come in contact with the soil as this supplies the additional iron that they require. Piglets in intensive commercial production suffer iron deficiency because they never come in contact with the soil and must be given an iron injection soon after birth.

Strict nutritional management is employed. Scientifically formulated rations maximise growth in relation to food intake. Growth rates are further optimised through the use of anti-microbial food additives. Disease spreads rapidly within an intensive commercial production unit, necessitating the use of antibiotics at the first sign of disease outbreaks.

The lifespan of a boar in an intensive factory system is eighteen months to two years. Sows have a productive life of only four or five litters or around two to three years under intensive production. Sows have lower offspring rates and reduced milk production beyond three years of age. Commercial producers intent on maximising production cull sows beyond this age.

Free-range Production

Foraging is a natural activity for pigs. Fresh air, exercise and sunlight are important for the good health of all animals. Free ranging your pigs has several advantages. Free-range pigs can graze at will. Even if the pasture provides little food value in the nutritional sense, it provides bulk that fills the pig's stomach and gives it a feeling of satisfaction. Pigs are more content, less noisy and have fewer digestive problems when allowed access to pasture. A large part of the pig's diet can be provided by planting a well planned forage system.

Free-range pigs exercise themselves. Pens require cleaning less frequently as given the choice, the pigs will defecate outside.

Being hard-hoofed animals, pigs do bring about compaction and degradation of land, particularly on steep slopes. However, pigs can be quite happily housed in pens with concrete floors, and can be allowed access to improved pasture for a few hours each day, thereby reducing land degradation.

Many books advocate placing a ring through the nose of free-ranging pigs, to stop them digging up the ground with their snouts. This is by all account a painless process, but try telling the pigs that! We have not found it necessary to ring our pigs, but keep paddock damage to a minimum by rotating animals at regular intervals. Trees should be fenced off from pigs to prevent them digging up the roots. If you intend to ring your pigs, the procedure is best done when piglets are young and under the guidance of an experienced person.

8. Feeding

All pigs are omnivores and as such are able to eat a great variety of foods. Pigs have small stomachs and need to be fed twice per day, once in the morning and once in the evening.

In a well designed, productive, free-ranging system it may be possible to reduce this to one feed per day, providing the animals have adequate access to a variety of free-range foods. Without adequate grazing, a once a day feeding regimen is likely to cause increased digestive problems among animals, since they tend to become more anxious at feed times and gorge themselves when food is given. Most pigs have a limited ability to ferment cellulose from plants and therefore cannot be sustained on pasture grazing alone.

You should aim to keep your pigs fit, active and not overfed. Despite common perceptions, pigs are not naturally fat animals. Overfeeding pigs results in excessively sized animals. Boars can become lazy and less interested in sows. Sows may experience difficulty in becoming pregnant or bear litters with only small numbers of piglets.

Commercially Prepared Feeds

When using commercially prepared pellets or meal, calculating the feed requirements of individual pigs is an easy exercise. Pigs usually require around two kilograms of pellets or meal per day, divided between the morning and evening feeds. Lactating sows require an additional half kilogram per piglet or can be fed a lactating sow diet ad lib. You do not need to weigh the grain each time, but can simply select a container (we use an old saucepan) that holds the required ration.

While feeding pigs scientifically formulated commercially prepared food certainly makes life easy for the pig keeper, they are expensive. Small-scale producers who rely solely on such products will find that it costs almost as much to raise their pigs as it does to buy the meat directly from the butcher. Some people also regard the sole use of such products as unsustainable as the manufacture of commercially prepared foods uses up valuable non-renewable resources.

Cost of raising a pig to porker weight on commercial pig rations

Age	Amount consumed	Cost
0-3 weeks	-	-
4-6 weeks	Average 250 g day	$ 1.96
7-16 weeks	Average 2 kg day	$52.50
Total		**$54.46**

Based on cost of $15.00 per 40 kg bag (37.5 cents per kgs) of pelleted food. Pig should reach porker weight of 60 - 70 kilograms live weight by 16 weeks. This amount does not include cost of maintaining the sow, maintaining/accessing a boar or AI, butchers costs and transport. Pig meal, when available, is often cheaper than pelleted products.

Commercially prepared pig rations vary in their components depending on the time of year, cost and availability of products. Some food products contain anti-microbial compounds designed to stabilise gut flora. Read labels carefully and find a product that fits your requirements. This is essential for those wanting to avoid unintentionally feeding their animals these additives.

Even well planned, free-range systems may need to use some commercially prepared feeds or grain to supplement the diet of pigs at different times of the year.

Free-range pigs develop less saturated fats. This is not a problem when porkers are being produced, in fact some would say that this is healthier for those consuming the meat. This fat needs to be hardened in pigs to be used for ham and bacon production. This can be achieved by feeding grains such as cracked corn or commercially prepared food as a proportion of the diet a week or two before slaughter.

Alternative Food Sources

Swill feeding of pigs, that is, the collection of all manner of foodstuffs and boiling them for consumption by pigs, is now illegal. Similarly, meat products cannot be legally fed to pigs due to the possibility of spreading serious diseases. However, recycling waste food products from bakeries, greengrocers, biscuit or cereal factories, vegetarian restaurants or other outlets can be a great way of reducing the need to buy commercially prepared food or grain supplements for your pigs. The growth rate of pigs fed on such by-products may be slower that those fed commercially prepared rations. However, if this material can be obtained at little or no cost, the fact that the pigs must be fed for a longer period is of little consequence to the small-scale pig keeper.

If you drive to and from work, look for outlets that may supply you with suitable waste products. Talk to your local suppliers. Remember that they are doing you a favour, so you will need to make it as easy as possible for them to provide you with waste materials. Arrange regular collection days. Provide them with specially marked bins in which to place the material. Keep the bins and surrounding area clean and tidy.

If you find that local suppliers already recycle their waste through other avenues, think laterally. We live in a semi-rural area and the products we want for our pigs are also consumed by a variety of other animals. We found all our possible local supplies had been spoken for. However, when travelling to work we drive through some up-market suburban areas. Demand for waste products is often non-existent in such areas. Few residents keep animals and those that do are not in the habit of collecting food from local bakeries or the greengrocers.

Growing Your Own

Like most people, we were under the impression that pigs would eat anything. We madly planted fruiting trees and vines for our pigs. Acknowledging that these would take several years to come into production, we grew plenty of annual and perennial crops for the pigs to eat.

What all our reading did not tell us was what our pigs would not eat. Our egg plants and okra produced bumper crops, but the pigs would not touch them. Capsicums were greeted with similar distain. Through trial and lots of error we discovered that given the choice, pigs have definite food preferences. Sweet fruits such as grapes and kiwi fruit are preferred over pumpkins and cucumbers. These are preferable to carrots and sweet potatoes. So far as our pigs are concerned, anything is better than raw capsicum.

Most foods will be eaten if cooked. Cooking pig food is generally time consuming and uneconomical unless you use a slow combustion stove for heating or cooking and can let a large pot simmer away with little effort or expense.

We have found boars to be less fussy than sows, probably because they always seem to be hungry. We try to introduce as many foods as possible to our pigs at an early age, that way they are more likely to accept a variety of foods.

The following information has been compiled by offering groups of various foods to our pigs and observing their preferences. While food preferences vary from one animal to the next, this information may provide a guide as to the proportion of each that should be planted. Consider the harvest period of these crops before planting if you wish to have supplies available throughout the year.

Food Preferences

Preferred Foods

Fruits

Apple, Apricot, Avocado, Babaco, Banana (peeled), Cucumber, Dates, Feijoa, Figs, Grape, Kiwifruit, Mango, Melon, Mulberry (fruits and leaves), Nectarine, Panama Berry, Pawpaw, Peach, Pear, Plum, Rosella, Strawberry.

Vegetables

Pea (pods and greens), Potato (cooked), Pumpkin, Sweet Potato (raw tubers and fresh greens).

Other

Barley (cooked), Biscuits, Bread, Cake, Clover, Corn (fresh or cracked), Comfrey (leaves), Eggs, Madeira Vine, Milk (and milk byproducts), Nuts, Lollies, Lucerne (fresh), Oats (cooked or ground), Pig Weed, Processed Cereals, Wheat (cooked or ground).

Eaten when Hungry

Fruits

Carambola, Custard Apple, Loquat, Passionfruit (pulp), Pomegranate, Tomato (over-ripe).

Vegetables

Asparagus, Bean (pods and greens), Beetroot, Broccoli, Brussels Sprout, Cabbage, Carrot, Cauliflower, Jerusalem Artichoke (cooked), Kale, Kohl Rabi, Lettuce, Potato (uncooked), Swede, Squash, Turnip, Zucchini.

Other

Bracken, Canna species, Cobbler's Pegs, Lucerne (dried), Tagasaste, Pigeon Pea (foliage and seeds).

When there is Nothing Else

Fruits

Capsicum (with reluctance when cooked), Citrus (unless over-ripe and cut in half, peel is not eaten), Granadilla, Egg Plant (with reluctance when cooked), Onion, Persimmon, Pineapple (unless over-ripe and cut in half), Tamarillo, Tomato (green).

Vegetables

Celery, Choko (preferably cooked), Okra (not even when cooked), Radish, Silverbeet.

If you have a variety of foods available for your pigs, give them the less favoured items first. They can be encouraged to eat these when they are very hungry, but will reject them if the edge is taken off their appetite after having eaten the preferred items first.

Interestingly, some plants suggested in books as pig forage are not favoured by our pigs. We planted dozens of pigeon pea and leucaena only to find that our pigs were not that keen on them. We have since discovered that pigeon pea is a good winter supplement. The pigs will consume the foliage and relish the developing seed pods in winter when pasture growth is slow. We have also been able to encourage our pigs to eat prunings of some tree species by cutting them up into pieces about ten centimetres long. A friend suggested that our pigs are just too well fed. She is probably right!

Food for your pigs can come from some strange sources. Madiera vine is a terrible bushland weed in our area, yet our pigs love it. It is so invasive that we would never suggest actually planting it for pig food, but a heavily infested area of bushland close to us provides a readily accessible and seemingly endless source of greens.

Grazing

Pigs will selectively graze pastures. Our pigs prefer fresh couch grass, rye, paspalum and pig weed, to other grass or weed species on unimproved paddocks.

Pasture can be improved by the inclusion of species such as clover, lucerne and millet. The species appropriate to different areas vary according to soil type, rainfall and seasonal temperatures. Agricultural departments produce information leaflets on legume and pasture species for different regions. Some suggested species for pasture improvement and orchard planting are listed on page 29.

While some pig keepers advocate using comfrey, chicory and arrowroot within pasture grazing systems, we have found it easier to grow these plants outside the grazing area and simply throw them over the fence. We have found that our pigs relish these alternative food sources so much that if left to graze they just about eradicate them from the paddock.

Orchards

Developing forage orchards for free-range pigs is advocated by some free-range pig keepers, since it is less labour intensive, allows the pigs to feed at will and provides a variety of food sources throughout the year.

If this is to be done it is important to establish the trees for at least five years before introducing your pigs. The roots and trunks of the trees should be protected by secure fencing. A variety of pasture legumes can be established as permanent ground covers beneath the trees.

We have planted our fruit trees, vegetables and root crops outside the forage area but close to the pigs. We

Species for pasture improvement and orchard planting

Warm Climates

Sown in warmer months

Glen Joint Vetch
Crown Vetch
Pinto's or Amarillo peanut
Pigeon Pea (shrub legume)
Sorghum
Cowpea
Japanese Millet
Lab Lab
Shaw Creeping Vigna

Sown in cooler months

White Clover
Strawberry Clover
Red Clover
Lucerne
Maku lotus

Cool Climates

Sown in warmer months

Buckwheat
Cowpea
Japanese Millet
Sorghum
Crown Vetch

Sown in cooler months

Lupins
Oats
Rapeseed
Strawberry Clover
White Clover

---

harvest or cut and carry fruits, vegetables and foliage to the pigs. While the orchard forage system has merit, it takes a long time to establish and trees can really receive a beating if a determined pig breaks through tree barriers. Food sources for our small number of pigs are provided by a mixed orchard and vegetable garden.

It is necessary to plan for year-round harvest if you wish to have an adequate supply of fresh fruits available for your pigs throughout the year. Listed on page 30 are some suggestions for subtropical climates. Stone fruits, nuts and acorns should be substituted in cooler areas.

Fruiting periods vary from one region to another and according to variety. Comfrey, sweet potato and arrowroot provide almost year-round harvest of greens that are relished by pigs.

Plants to Avoid

The plant species that are toxic to stock are too numerous to list here. Many

Suggested fruits for subtropical areas

Fruit Tree/Vine Crop	Fruit Harvest Period
Mango	December - January
Grapes	January - March
Figs	February - March
Feijoa	February - March
Rosella	February - May
Custard Apple	March - August
Pecan	April - May
Avocado	August - November (Hass variety)
Tropical Peach	August - September (Florda varieties)
Bananas	October - March
Jaboticaba	October - November
Pawpaw	October - February (substitute Babaco in cooler areas)
Mulberry	November (foliage Summer & early Autumn)

excellent texts have been written on this topic (see bibliography). Most small landholders would not recognise many of the species. Fortunately, adult pigs seem to be intelligent enough to avoid the poisonous species or tough enough to tolerate the occasional nibble.

We once made the mistake of letting two young, adventurous pigs into a paddock in which we were growing some rhubarb. We knew that rhubarb leaves were poisonous and mistakenly assumed that the pigs would naturally avoid it. In fact, they devoured five crowns, leaves, stalks, roots and all.

We spent a few anxious hours watching for signs of listlessness or vomiting. We calculated that the cost of calling the vet out on a weekend would probably be more than the pigs were worth and that it was unlikely that they could do anything anyway. The pigs devoured their evening grain ration and ran around the paddock just as they always did. They showed no ill effects from their encounter with the rhubarb and while we would not recommend that you knowingly allow your pigs to graze on poisonous vegetation, it happens.

If you intend to allow your pigs to graze in paddocks and have two or three dominant weed species present, it may be worth getting these identified just in case they represent a hazard. Neighbours, produce stores, your local department of primary industries office or the vet can often provide free identification and advice.

Food Values

The food values of commercially prepared rations are stated clearly on the label. Ascertaining the food values of fruit, vegetables, breads and other possible food products is more difficult. It is possible to purchase individual components similar to those contained in commercial foods and produce your own pig feed formula. If components are purchased in bulk this is often cheaper.

Knowing what and how much to feed your pigs involves a little bit of trial and error. Books on modern pig production do not give you any idea of the food values of waste food or home-grown crops. Some useful information and interesting reading can be found in older books on pig rearing. Scour the secondhand bookshops for old English, Australian and New Zealand texts on pig husbandry. Books written between 1940 and 1955 often have tables outlining the food values of fruit, vegetables and various grains. They may even give details as to the proportion and quantities of each item required to give a well balanced diet.

Kelp or seaweed supplements are now recommended for a wide range of stock, including pigs. Liquid kelp contains vital dietary trace elements. It may be a useful inclusion in the diet of pigs not fed commercial pellets or meal, particularly where grazing pastures are of poor quality.

Coleby (1991) suggests that all pigs benefit from having the following dietary additives.

Dolomite - one teaspoon per head daily

Seaweed products of some kind - ad lib for preference

Sulphur - one teaspoon per head daily

Cod liver oil - one teaspoon once a week at least

In days gone by pigs were given access to wood ash and charcoal. The theory was that when eaten by the pig this material provided worm control because of its alkaline nature and grit in the cinders. As we have a wood fire we decided to try it. We really did not think the pigs would eat the wood ash, but they readily consumed it. We do not know whether it aids in worm control, but it no doubt adds lime (calcium) and potassium to their diet.

Listed below are a few basics we have learnt about feeding pigs. Most are just common sense.

1. Pigs are happiest and healthiest when fed a variety of fruits, vegetables, grains and forage greens. This provides them with a range of nutrients.

2. The proportion of their diet that can be supplemented by grazing varies depending on the type of pig and the

type and quality of pasture, root crops, fruit and nut fall provided. Even in a well planned grazing system it is difficult to supply all the food needs of the pig. You should expect to supplement this with fruit, vegetables, breads, cereals or commercially produced products.

3. While it is true that fat deposition in hybrid pigs is largely genetic (breeding stock is selected on the basis of low fat deposition), this characteristic has not been developed to the same extent with pure breeds and some management of their diet is required to minimise development of excess fat.

4. Pigs with access to grazing get more exercise and therefore are less likely to put on excess fat.

5. Pigs have a sweet tooth and although they relish biscuits, cakes and other sweet products, excess intake of these foods will tend to make them fat. If your pigs get too fat, gradually reduce their food intake, change the proportions of various dietary components and give them more exercise. Pigs fed poorly balanced diets may also suffer from skin conditions.

6. On average, pigs should take about twenty minutes to consume their food ration. Pigs should be encouraged to eat and grow quickly when they are young, but given maintenance rations as they mature beyond porker size. You must reduce the food intake of most purebreed pigs in the later stages of growth to prevent excess fat being formed.

7. Root crops and tubers such as potatoes have a higher protein value when fed to the pigs cooked as opposed to uncooked.

8. Lactating sows have higher food requirements, particularly in the first few weeks after farrowing. Underfed lactating sows loose condition quickly and piglets suffer from reduced milk production.

9. Green crops cut for feeding to pigs must be fresh. Wilted greens are not only less palatable, but have lost a proportion of their food value. Dried hay or grasses provide bulk, but little nutrition. Pigs fed a high proportion of fresh, over-ripe fruits will smell more than those fed a balanced diet.

10. Pigs use a significant proportion of their food intake maintaining body warmth in winter. They are likely to need slightly more feed in cooler weather.

Storing Foods

Recycled 200-litre drums are useful for storing food. Select clean drums with lids. You must be able to seal the drum well enough to keep out rats and mice. One drum will hold three forty-kilogram bags of commercial feed or can be used for dry foodstuffs such as bread and cereal. Drums should be stored under cover, so it is sensible to plan a food storage area when designing your pens.

If you are collecting waste fruit and vegetables, small amounts can be stored in an old refrigerator. Waste food deteriorates quickly in hot weather and becomes unpalatable to pigs. Select a cool, shaded, well ventilated area to store foodstuffs. Wet hessian or canvas can be used as a simple evaporative cooler around perishable greens.

Feeding Practicalities

Pigs are naturally inquisitive animals. When you enter the pen they will come up and nuzzle you wanting a scratch, some food or just to say hello. Even the cleanest pigs will slobber over your legs and clean clothes. Keep an oversized pair of overalls handy. These can be quickly pulled on at feeding time. Alternatively, arrange the layout of pens so that feed containers can be filled without actually entering the pen.

Pens are easier to clean and less food wastage occurs if food is placed in troughs or other suitable containers. If possible, these are best fixed in place. Pigs will play with moveable containers, tossing them around the pen or paddock. We purchased some semi-circular concrete drain pipes cheaply at an auction. They make excellent, durable feeding troughs. Lengths of guttering with the ends sealed off make great feeders too, especially for smaller pigs. Mount them on timber for extra durability. Cut down 200-litre drums, or tractor tyres cut in two are also cheap, effective feed troughs.

Make sure that you allow plenty of room for each pig to feed. We have a separate trough for each pig. This minimises squabbles over food. They soon learn which trough belongs to them and rush to it each feed time.

It is possible to construct a simple hopper system for ad lib feeding. This will enable you to go away for the odd weekend. Simply obtain a length of PVC or galvanised downpipe 100 to 150 millimetres in diameter and about one metre in length. Cut a section from one end of the pipe **as illustrated below.**

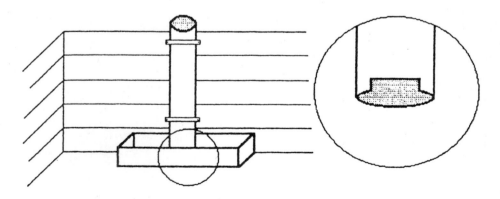

A simple hopper system for ad lib feeding.

Secure the length of pipe so that it sits into the feeding trough.

Fill the pipe with commercially prepared rations or other dry, flowable feed. This will gradually spill into the trough as the animal eats. The length and diameter of the pipe will determine the amount of feed that can be held and the length of time the pigs can be left unattended. This ad lib feeding system is also a useful way to ensure that young pigs have continuous access to feed and grow at the required rate during the first three to four months.

9. Watering

Like all animals, pigs need unlimited access to clean, fresh drinking water. A pig's food intake is to some extent limited by the availability of clean drinking water. When feeding pellets, meal, bread or cereals, pigs will often alternate between the water and the food trough as they feed.

Water is becoming an increasingly expensive commodity. If you are building any structures to house your pigs, it makes sense to consider collecting rainwater from the roof. Even those on mains supply should consider diverting water from the house roof into a tank for use by animals. We collect rainwater from the roof of our pig housing in a 7000-litre galvanised tank. This provides more that enough water for a boar, two sows and any offspring, while still allowing plenty for cleaning pens and washing the pigs.

Calculating the catchment area of your roof and working out the size tank you need is relatively easy. You will need to know the average monthly rainfalls for your area and the likely intervals between good rain periods. We store enough water for three months without rain. Adjust this to your local weather pattern.

Only 70% of possible runoff is calculated since approximately 30% of runoff is lost through water splash and evaporation. Use the equations below to calculate your optimum tank capacity.

Various containers are suitable for holding water for pigs. Ideally they should be fixed or at least tied to one corner of the pen or paddock to prevent them being tipped over. As animals often waste water from open troughs, it is difficult to ascertain exactly how much they drink. However, we estimate that our boar drinks 35 litres of water per day during the summer. A lactating sow rearing a large litter needs at least 45 litres per day. If using open containers, allow plenty of extra water for wastage.

Calculating catchment and tank capacity

Roof Area (m^2) x Average monthly rainfall (mm) x 0.70 = Average monthly catchment (litres)

Average monthly catchment (litres) x Number of months storage required = Optimum tank capacity

Water use in an intensive commercial piggery is incredible. An allowance of 225 litres per day is made for one sow and her offspring for consumption, washing and flushing of waste. Disposing of this runoff presents a major difficulty for commercial operations.

As we rely solely on tank water for our house, garden and animals, we have installed pig waterers or nipple drinkers. These are made from stainless steel and the pigs bite on them to release the water flow.

Using pig waterers has several advantages. Clean, fresh water is always available to the animals. Water wastage is minimised and considerably less time is involved in tending the animals since there is no need to wash and refill water containers.

One waterer will service several pigs. Pig waterers are mounted just above the head height of the pig. As you are likely to have pigs of various sizes using the same watering point, it is useful to mount pig waterers at different heights.

Pigs love to play in water. Sometimes it appears as if they are just bored and tip water containers over just for the fun of it. At other times they will want to cool down. Pigs do not sweat through their skin, but pant like dogs, breathing though their mouths when they are heat stressed.

Pigs wallow in mud to keep cool and rid themselves of parasites. Skin coated in mud is also less likely to burn. We have found pig wallows messy, smelly and unnecessary, providing pigs have plenty of shade and external parasites are controlled by other means.

Pigs appreciate a daily hosing in very hot weather or better still a bath. Dig a hole and sink an old bath into the ground to prevent the pigs tossing it around. Connect a piece of downpipe to the plug outlet and bury it under the ground with the outlet near productive trees. Each time you empty the bath you can use the waste water on the garden.

10. Housing

In mild climates, inexpensive ark-type housing can be used for free-range pigs. Such field huts need to be large enough to accommodate the required number of pigs, but sufficiently portable so that they can be moved from one paddock to the next when rotating animals. If you intend to allow sows to farrow in field huts, it is useful to provide some protection for piglets. Straw bales can be used to create a small run. A low wooden shroud that the sow is able to step over can also be used to keep young piglets in.

Temporary housing can be made from straw bales stacked one on top of another. About twelve bales are required. Make sure they are well bound and tie extra binding twine around them to stop the straw falling out. Drive a few star pickets through the bales to secure them in place. Cover the top with iron sheets well anchored to the star pickets or the ground.

For more permanent housing, we believe each sow or boar requires a covered area of at least 2.5 x 2.5 metres, plus an adjacent outdoor exercise area 2.5 x 3 metres. This is extravagant by intensive production standards. Intensive production units on average allow just over two square metres per mature pig.

Be sure you have a clear idea of how large your particular breed of pig will grow. We really had no idea that Tamworth boars and mature sows would grow to the size they do! We initially made our pens 2.5 x 2.5 metres providing what we thought was a generous increase on recommendations we had read in books. This may be suitable for overnight housing, or short periods of confinement. However, during long periods of wet weather our pigs are confined to their quarters. They love to dig up wet ground and can catch colds and pneumonia easily under wet conditions. Providing a sleeping, eating and exercise area totalling 2.5 metres wide and 5.5 metres long accommodates a large boar, a sow and her litter or several young boars or gilts in comfort. Larger pens are also easier to clean, result in less fighting among pigs and provide more humane conditions during periods of confinement.

Housing should allow morning sun to enter the pens, but provide protection from hot afternoon sun, especially in warm climates. Housing should be sufficiently ventilated in summer, but keep pigs free from cold draughts and driving rain. We have a roof over the 2.5 x 2.5 section of the pen, while the exercise area remains uncovered so that the pigs can bask in the sun occasionally.

It is possible to start off with simple inexpensive huts or adapt existing structures. Housing can be upgraded as demand arises and materials become available. We managed to construct quite substantial pig accommodation from waste materials left over from the construction of our house.

Choose a housing style suited to your climate and the number of pigs you intend to keep. Plan a layout that will work on your site, taking into account the maximum number of pigs you are likely to have at any one time. Design a system whereby additional pens can be added later on. You always end up keeping more pigs than you originally intended. It is also useful to have space to look after other people's pigs, since a reciprocal arrangement can be set up when you want to go away on holiday.

We can accommodate one boar, two sows and their offspring in pens detailed in the diagram. An additional gate on one pen enables a farrowing sow to be isolated. A boar and sow can be housed comfortably in the remaining pen. A series of gates allows a rotational system of access to paddocks. Drainage and any manure that washes from the pens runs

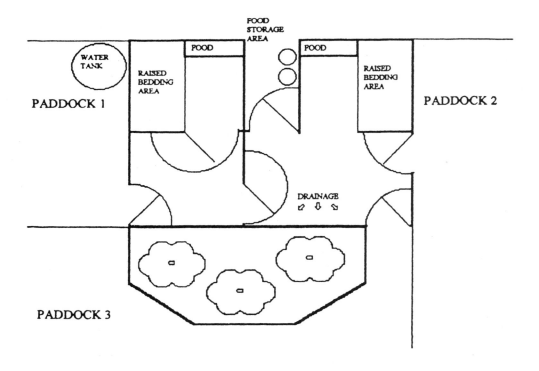

on to a garden bed below. This has been planted with fruiting trees that provide shelter for the pens and food for the pigs.

Concrete or similar flooring materials are durable and easy to clean. Make sure that they are not too smooth or your pigs will slip over. We provide a slightly raised, wooden floor area within the pens for our pigs to sleep. Alternatively, you can spread hay, sawdust or other bedding materials in one corner.

It is best to try to enclose pens partially with solid walls or at least screen the pens so that pigs cannot see you. Some pigs think every time you make a move you are coming to feed them. They will squeal in frustration if the anticipated food does not arrive. Obscure their view of you and your family, but use wire or wooden slats as dividing walls between the pens so that pigs can see each other.

11. Fencing

Free ranging your pigs can mean additional costs in terms of fencing requirements. Pigs are powerful animals and require strong fencing. As pigs are inquisitive by nature, they are excellent escape artists. Pigs will also rub up against fences or strain through the wire for that elusive piece of grass on the other side.

We decided to set up permanent fence lines and found pig wire to be the easiest and most successful product to use. Pig wire consists of hinged jointed rectangles with line wire spalings 7.5 cm apart at the base increasing to 15 cm at the top. Pig wire is 80 cm high and comes in rolls 100 m long. We have since discovered that goat wire is similar, only slightly more widely spaced. It comes in rolls of 200 m and is considerably cheaper. Pig wire fencing is quick and easy to erect using star pickets or wooden posts at two-metre intervals. One row of barbed wire along the base of the fence will stop the pigs from lifting the fence or digging out underneath it. An additional row of barb wire positioned 10 cm above the top of the pig wire can be used to prevent pigs from climbing over the top. Having said that, our boar recently forced his way between the top barb and the pig wire in a determined effort to reach a sow on heat.

Two-strand electric fencing is a cheaper alternative to permanent fencing. Pigs are intelligent and quickly learn where their boundaries are. Electric fencing is relatively cheap, easy to move and can be connected to batteries or mains power. You could even set up a system using solar power and storage batteries.

If fencing already exists it may be possible to adapt it for pigs using electric fencing or by attaching additional barbed wire. It may also be useful to plant living fences of shrubs that are unattractive to the pigs along less secure fencing. Thorny plants such as pyracantha and berberis are suggested.

12. Breeding

Young gilts and boars running together are often stimulated into early reproductive maturity. Pigs may reach breeding age as young as five months. However, it is preferable to wait until at least seven months of age before allowing sows to mate and seven to eight months for boars.

Sows come into heat approximately every 21 days and can be successfully mated for about three days. Sows in season will have a discharge and will stand ready to be mated when you lean heavily on the rear of their back.

Boars will show interest in sows several days before they are ready to be mated. Be prepared for the ensuing chasing and squeals. Mating of pigs is generally a noisy affair!

Planning Production

With a little planning it is possible to time the birth and development of litters to coincide with periods of peak demand. The gestation chart provided, (pages 42 and 43), is based on a 114 day gestation period. By recording the date of mating it is possible to ascertain the approximate farrowing date. Calculations can then be made as to when pigs will have reached pork and ham/bacon weights.

When planning for periods of peak demand such as Christmas and Easter, keep in mind that this is also a busy time for abattoirs and butchers. Be sure to telephone in advance and book your animals in. Pork is normally ready three to five days after slaughter, whereas hams and bacon generally require a further seven to ten days for curing.

Pigs mated on 21 April will farrow on 12 August and should reach porker weight by 2 December (16 weeks) just in time for Christmas.

Pigs mated on 10 February will farrow on 3 June and should reach ham/baconer weight by 2 December (26 weeks).

Matching Sows and Boars

It is important to match the size of mating sows and boars. Young sows may collapse under the weight of a very large boar. The experience will stress both animals and may make the sow reluctant to mate in future. Equally, a small boar will not be able to mate with a large sow. The eager boar may find the large sow unreceptive to his advances. She may become aggressive and he is likely to come off second best.

GESTATION CHART

January	Farrowing Date	February	Farrowing Date	March	Farrowing Date	April	Farrowing Date	May	Farrowing Date	June	Farrowing Date
1	24 April	1	25 May	1	22 June	1	23 July	1	22 August	1	22 September
2	25 April	2	26 May	2	23 June	2	24 July	2	23 August	2	23 September
3	26 April	3	27 May	3	24 June	3	25 July	3	24 August	3	24 September
4	27 April	4	28 May	4	25 June	4	26 July	4	25 August	4	25 September
5	28 April	5	29 May	5	26 June	5	27 July	5	26 August	5	26 September
6	29 April	6	30 May	6	27 June	6	28 July	6	27 August	6	27 September
7	30 April	7	31 May	7	28 June	7	29 July	7	28 August	7	28 September
8	1 May	8	1 June	8	29 June	8	30 July	8	29 August	8	29 September
9	2 May	9	2 June	9	30 June	9	31 July	9	30 August	9	30 September
10	3 May	10	3 June	10	1 July	10	1 August	10	31 August	10	1 October
11	4 May	11	4 June	11	2 July	11	2 August	11	1 September	11	2 October
12	5 May	12	5 June	12	3 July	12	3 August	12	2 September	12	3 October
13	6 May	13	6 June	13	4 July	13	4 August	13	3 September	13	4 October
14	7 May	14	7 June	14	5 July	14	5 August	14	4 September	14	5 October
15	8 May	15	8 June	15	6 July	15	6 August	15	5 September	15	6 October
16	9 May	16	9 June	16	7 July	16	7 August	16	6 September	16	7 October
17	10 May	17	10 June	17	8 July	17	8 August	17	7 September	17	8 October
18	11 May	18	11 June	18	9 July	18	9 August	18	8 September	18	9 October
19	12 May	19	12 June	19	10 July	19	10 August	19	9 September	19	10 October
20	13 May	20	13 June	20	11 July	20	11 August	20	10 September	20	11 October
21	14 May	21	14 June	21	12 July	21	12 August	21	11 September	21	12 October
22	15 May	22	15 June	22	13 July	22	13 August	22	12 September	22	13 October
23	16 May	23	16 June	23	14 July	23	14 August	23	13 September	23	14 October
24	17 May	24	17 June	24	15 July	24	15 August	24	14 September	24	15 October
25	18 May	25	18 June	25	16 July	25	16 August	25	15 September	25	16 October
26	19 May	26	19 June	26	17 July	26	17 August	26	16 September	26	17 October
27	20 May	27	20 June	27	18 July	27	18 August	27	17 September	27	18 October
28	21 May	28	21 June	28	19 July	28	19 August	28	18 September	28	19 October
29	22 May	—	—	29	20 July	29	20 August	29	19 September	29	20 October
30	23 May	—	—	30	21 July	30	21 August	30	20 September	30	21 October
31	24 May	—	—	31	22 July	—	—	31	21 September	—	—

July	Farrowing Date	August	Farrowing Date	September	Farrowing Date	October	Farrowing Date	November	Farrowing Date	December	Farrowing Date
1	22 October	1	22 November	1	23 December	1	22 January	1	22 February	1	24 March
2	23 October	2	23 November	2	24 December	2	23 January	2	23 February	2	25 March
3	24 October	3	24 November	3	25 December	3	24 January	3	24 February	3	26 March
4	25 October	4	25 November	4	26 December	4	25 January	4	25 February	4	27 March
5	26 October	5	26 November	5	27 December	5	26 January	5	26 February	5	28 March
6	27 October	6	27 November	6	28 December	6	27 January	6	27 February	6	29 March
7	28 October	7	28 November	7	29 December	7	28 January	7	28 February	7	30 March
8	29 October	8	29 November	8	30 December	8	29 January	8	1 March	8	31 March
9	30 October	9	30 November	9	31 December	9	30 January	9	2 March	9	1 April
10	31 October	10	1 December	10	1 January	10	31 January	10	3 March	10	2 April
11	1 November	11	2 December	11	2 January	11	1 February	11	4 March	11	3 April
12	2 November	12	3 December	12	3 January	12	2 February	12	5 March	12	4 April
13	3 November	13	4 December	13	4 January	13	3 February	13	6 March	13	5 April
14	4 November	14	5 December	14	5 January	14	4 February	14	7 March	14	6 April
15	5 November	15	6 December	15	6 January	15	5 February	15	8 March	15	7 April
16	6 November	16	7 December	16	7 January	16	6 February	16	9 March	16	8 April
17	7 November	17	8 December	17	8 January	17	7 February	17	10 March	17	9 April
18	8 November	18	9 December	18	9 January	18	8 February	18	11 March	18	10 April
19	9 November	19	10 December	19	10 January	19	9 February	19	12 March	19	11 April
20	10 November	20	11 December	20	11 January	20	10 February	20	13 March	20	12 April
21	11 November	21	12 December	21	12 January	21	11 February	21	14 March	21	13 April
22	12 November	22	13 December	22	13 January	22	12 February	22	15 March	22	14 April
23	13 November	23	14 December	23	14 January	23	13 February	23	16 March	23	15 April
24	14 November	24	15 December	24	15 January	24	14 February	24	17 March	24	16 April
25	15 November	25	16 December	25	16 January	25	15 February	25	18 March	25	17 April
26	16 November	26	17 December	26	17 January	26	16 February	26	19 March	26	18 April
27	17 November	27	18 December	27	18 January	27	17 February	27	20 March	27	19 April
28	18 November	28	19 December	28	19 January	28	18 February	28	21 March	28	20 April
29	19 November	29	20 December	29	20 January	29	19 February	29	22 March	29	21 April
30	20 November	30	21 December	30	21 January	30	20 February	30	23 March	30	22 April
31	21 November	31	22 December	—		31	21 February	—		31	23 April

Preparing for Mating

Mating pairs are best confined to a large run or a small paddock. Heat-stressed boars will be less fertile. Try to provide cool conditions, particularly shade and plenty of water. Only allow one on-heat sow with a boar at any one time, or the boar is likely to exhaust himself. This may result in smaller litters.

If you do not own a boar you must take your sow to be mated. This is preferable to taking the boar to the sow. If the owner of the boar is agreeable, try to arrange for delivery of your sow about one week before she is due to come into season. This will allow the sow time to settle and overcome any stress associated with the journey. Some pigs get travel sickness and will be off their food for a day for two after relocation.

The services of the boar should not be expected free of charge. Considerable expense is involved in feeding, housing and maintaining a boar in good condition. Servicing of sows is the one way that the boar can pay his way. You should expect to pay a service fee for the efforts of the boar, plus supply or reimburse the cost of feeding your sow during the mating period. Some boar owners are happy to barter. One offspring from the resulting litter at porker weight would be an acceptable payment. If there are several sow owners in your area you may decide to keep a single boar, plus the offspring bartered from service fees. Commercially, a working boar is expected to service several sows in one week. In free-ranging systems the ratio of sows to boars is around twelve to one.

Artificial insemination (AI) is an alternative for those without access to the services of a boar. Semen is usually available from government agencies. Current cost is around $60 per sow. This cost is much lower than maintaining a boar when only a small number of sows are kept. Using AI is also one way of introducing new bloodlines or particular breed characteristics into your animals.

The process of artificially inseminating animals is not difficult for those with some experience in dealing with stock, however the timing is critical. It is essential to keep accurate records so you will know in advance when a sow is likely to come into season. The sow will only be receptive for about three days and several inseminations are usual during mid cycle. Semen is usually purchased just prior to use and you may need to travel a considerable distance to access supplies. Small-scale pig keepers contemplating use of AI need to be well organised.

Pigs have a gestation period of 114 to 116 days or three months, three weeks and three days. Pregnant sows should not be stressed in the weeks following mating as this can increase the risk of death in offspring. Transport animals in the cool of the day and try to minimise the distress associated with loading and unloading following mating.

It is tempting to give a pregnant sow extra food, after all she may be eating for ten. However, pregnant sows should not have rations increased substantially or they will become too fat. This may present problems during birth. Pregnant sows often suffer from constipation, particularly in the final weeks of gestation. The problem should be treated immediately by adding a little cooking or paraffin oil to their daily food rations. Sows in poor condition or pregnant during very cold weather may require increased food rations.

Preparing for Birth

If you have made a note of the mating time, you will be able to calculate the approximate date your sow is to give birth. Where possible, separate the sow from other pigs two to three weeks before farrowing. Make appropriate preparations by cleaning and disinfecting the farrowing pen one week before the sow is due.

A strong farrowing rail should be fitted to the pen to provide a safe area for piglets to shelter when the sow lays down. This will take the form of a piece of timber attached diagonally across one corner of the pen, 10 to 20 cm above floor level.

Before farrowing, the sow should be given a good wash using a mild soap. Laundry bar soap or anything you would wash your own hair or hands with is generally acceptable. Mites and other external parasites can be transferred from the sow to her piglets. The farrowing quarters and the sow should be completely free of any mite infestation before farrowing. If external parasites are suspected, wash the sow using a mild disinfectant or pyrethrum-based rinse. Pay particular attention to the ears and neck of the sow.

Confine the sow to a small run and the farrowing pen. Supply clean straw or sawdust as bedding material and allow the sow to become comfortable with her surroundings. Like most animals,

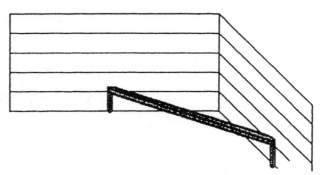

Farrowing rail diagram

a sow will look for somewhere safe and protected to have her young. If this is not provided, the sow will become very agitated.

As the farrowing time approaches the sow will begin making a nest using the bedding material provided. She will scrape the material with one hoof, then the other and push it around with her snout. Where straw is provided she may even chew it up into small pieces. We prefer to use sawdust as young pigs can become strangled by long lengths of straw.

In some sows, milk can be encouraged from the teats 12 hours before the sow gives birth. This is not always a reliable indicator as other sows will not let down their milk until several hours after giving birth. Nest making activity is a sure sign that the birth is not far away. If you know your sow is about to give birth, a small bottle of beer poured into a bucket can be given as a mild, oral analgesic.

Most sows give birth at night. The whole event may be over in just a few hours with piglets popping out faster than you can count them. In other instances it may take six to eight hours. Don't panic if things seem to be going a bit slowly. With pigs it is certainly a case of mother nature knowing best. We have even known sows to get up for a walk around or to finish their leftover dinner in the middle of farrowing. If you try to interfere with a farrowing pig in an attempt to hurry the process along you are likely to cause more problems than you solve.

The appearance of the afterbirth generally signals that farrowing is completed, although it is not unusual for some afterbirth to be expelled along with the piglets.

Some sows are happy to have familiar people in the pen with them right through the farrowing process. While this is not essential, it is sometimes useful to move piglets to a safe corner of the pen while farrowing continues. Some piglets can be encouraged to survive by clearing mucus from their mouth and blowing air gently into their lungs.

Some sows are very protective of their young and will not entertain anyone entering their domain during or following the birth of their young. To attempt to do so could put yourself and the piglets in danger and cause undue stress on the sow. In such cases it is especially important to make sure that all preparations previously detailed have been carried out in advance and that a secure farrowing rail is in place to provide some protection for the piglets from overlaying by the sow.

Gilts giving birth for the first time are often shocked by their new arrivals. It is common for gilts to initially reject piglets as something foreign and certainly not belonging to them. Do not try to force the sow to accept the piglets. She will soon become accustomed to them and allow them to suckle. In rare cases gilts have been known to cannibalise piglets.

Litter sizes vary, but eight piglets is a good average. Gilts tend to bear smaller-sized offspring and fewer piglets. Productivity increases with further litters and declines as the sow matures. Sows in non-intensive systems can bear good litters for six years or more.

New piglets are remarkably sturdy animals. They are clean, warm and soft as silk. As they are born they struggle to their feet and search out the best teat. Teats located at the front of the animal produce the most milk. The largest and strongest pigs are often born first and take their place instinctively on the first teats.

During the hours following birth the piglets jostle for the position. Once the hierarchy has been established each pig will return to its own teat for each feed. The runt of the litter is often the last piglet born and is consequently the smallest. It is also weaker than the other piglets and consigned to one of the lower, less productive teats.

Caring for Piglets

It is sometimes possible to encourage weaker piglets to drink by placing them on teats or providing them with supplementary feeding. Equal quantities of cow's milk and water to which a teaspoon of glucose or an egg has been added is suitable. Use a sterilised baby's bottle and feed the piglet very small amounts at least every four hours. It is important that all piglets obtain the colostrum provided by initial suckling from the sow since this passes on disease immunity.

Bottle feeding should only be used as a supplementary measure or as a last resort with an oversized litter or particularly weak piglet. The novelty of getting up every four hours to feed a piglet soon wears off. In our experience, animals that are solely bottle fed do not grow at the same rate or have the vigour of piglets raised by the sow.

Young piglets are very cute and you will not tire of watching them. They must be kept warm, dry and protected from draughts, particularly during cold or wet weather. Young piglets will huddle together to maintain their body temperature. In cool areas or during winter, a light globe suspended over the corner of the pen above the farrowing rail will help keep them warm. The air temperature should ideally be maintained just above 21 degrees Celsius. Piglets that are cold will be weak and less able to suckle. They will be restless and jostle for position within the middle of the huddle, using all their energy just trying to keep warm. They may also tend to stay close to the sow to take advantage of her body heat and are much more likely to be rolled on.

Commercial piggeries place lactating sows in farrowing crates. This reduces the incidence of overlay and restricts movement of the sow so that piglets can gain easy access to the teats at all times. We have found use of such devices unnecessary. Some sows seem very clumsy, while others go to great lengths, honking and grunting to give piglets warning of when they intend to lay down.

Good mothering is an inherited trait and sows should be selected for this as much as any other characteristic.

Our Tamworth boar is from the Ranger bloodline. At just 18 months old and weighing in excess of 200 kilograms, he is an incredibly powerful animal.

Digging for roots and tubers is a natural activity for pigs. They can be used to effectively clear areas infested with shrubby weeds.

Pigs enjoy a variety of foods. Mulberry leaves are a particular favourite.

Even piglets enjoy access to pasture. These robust youngsters are just eight weeks old.

Like many animals, pigs relish a cool hose down on a hot day.

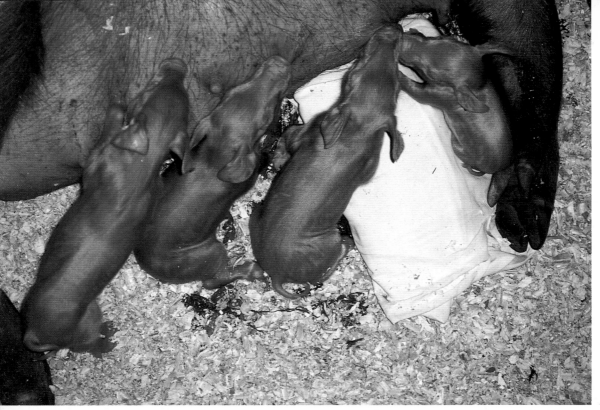

These newborn piglets are barely an hour old. Some sows are happy for you to be present throughout the farrowing process.

Pigs get along with most animals. Our kelpie loves to play with the young pigs, but has a healthy respect for the larger sows and our boar.

Pigs love to wallow in mud to cool down on a hot day. This also helps to protect their skin from sunburn and keep them free of external parasites.

It is important to provide cool, shaded areas for pigs at all times.

Pigs enjoy a good wash. Any mild soap is suitable, just be sure to rinse off well.

Modern hybrid pigs are popular with intensive commercial producers because of their phenomenal growth rate and prolific reproduction.

New Zealand has its own peculiar breed of pig. The Kunekune is thought to have been brought to New Zealand by the Maoris.

Transporting pigs need not involve great expense. We can easily accommodate several pigs on the back of our utility.

Pigs are very inquisitive animals. As the pens are being cleaned, these piglets take the opportunity to play with a running hose.

13. Growth and Development

The growth rate and development of pigs varies according to breed, climatic conditions, housing and feeding regimen. As a guide it will take sixteen to eighteen weeks or 4.5 months for a pig to reach porker weight of 70 kg. It will take twenty-six weeks or 6.5 months for a pig to reach the 90 kilograms required for ham or bacon.

We like to sell a couple of suckers from each litter. This reduces the number of pigs at the more demanding porker/baconer stage when it is difficult to accommodate eight to ten semi-mature pigs. This is also a good way of reducing boar numbers. We generally use our boars (other than breeding stock) for pork production. Slow maturing boars may reach sexual maturity, giving their meat a strong taint if they are grown on uncastrated past four to five months of age.

Piglets will suckle the sow for six to eight weeks. It is possible to remove them from the sow at four weeks if you wish to maximise production. The sow will come back into season less than a week after the piglets are removed. If you do not remove the piglets the sow will gradually wean them herself at between eight and ten weeks. Leaving the piglets on the sow is one way of delaying oestrus and can be used as a means of planning production to suit peak demand.

Piglets generally begin to show interest in solid foods at around three weeks. Scattering a little food on the floor for them will enhance their natural curiosity. Piglets gradually introduced to solid food while still suckling are less subject to scouring. This tends to be a problem when piglets are removed too early from the sow and/or when they are transferred to a solid diet.

Some books advocate the use of specially formulated creep feeds for young piglets. We have found this expensive and unnecessary. When piglets begin to show an interest in solids

Growth stage	Age in weeks	Weight in kilograms
Sucker	Birth - 6	1 - 10
Weaner	6 - 12+	10 - 30
Porker	12 - 18+	30 - 70
Baconer	18 - 26+	70 - 90+

we allow them access to food away from the sow. This is done by isolating them in a pen with a small amount of food while the sow is feeding. Saving a few tasty morsels for piglets will encourage them to feed. We keep hens and always have plenty of eggs. Piglets relish a couple of eggs mixed with their dry food ration. Soft or sweet fruits such as papaws, grapes and melons are also devoured.

It is important to ensure that pigs grow rapidly during their first 16 weeks of development. During this time they tend to gain weight in the form of flesh rather than fat. From this time on their food intake should be carefully monitored.

Measuring Growth

You will need to weigh your pigs to determine whether they are ready to go to market and in order to calculate dosage rates for worming preparations if you decide to use them.

How on earth do you weight a pig? They will not fit on the bathroom scales and few people have the sophisticated weighing crates used by commercial growers. Fortunately you can determine the weight of a pig by measuring the girth just behind the forelegs. Special live weight tapes are available, but a cheap and adequate substitute can be made using an old dressmaking tape measure or any similar washable flat tape that can be written on. Make up your own live weight tape using the following measurements.

Girth measurement	Approximate live weight
60cm	37 kilograms
65cm	42 kilograms
70cm	47 kilograms
75cm	53 kilograms
80cm	58 kilograms
85cm	63 kilograms
90cm	70 kilograms
95cm	75 kilograms
100cm	85 kilograms
105cm	95 kilograms

14. Handling and Management

Pigs respond to affection and are easier to handle if you treat them with kindness from an early age. Like humans, pigs vary in disposition. Pigs that have been mistreated can be aggressive and difficult to handle. Some sows have a tendency to be affectionate when in pig, but are tempted to give you a nip on the ankle at other times. Pigs are often large and powerful, so you cannot afford to keep a bad-tempered animal. This is particularly important with boars. If pigs are to be kept in good health you must be able to enter their pens safely for feeding and cleaning. It will be necessary at times to wash them and treat minor wounds and other injuries. Encourage your pigs to be docile by talking to them and handling them at an early age.

In common with most other animals, groups of pigs have their own hierarchy. Introducing a new pig to an established group results in a reshuffling of positions. The new pig can take quite a beating in the process. When introducing new pigs, allow them plenty of room. At least that way threatened pigs can get away from their attackers. Watch for any individual that seems to be particularly distressed. Masking identifying body odours of all pigs in the group with a strongly fragrant wash can minimise fighting. It takes the pigs a day or two to work out who's who!

Moving Pigs

If a pig does not want to go in the direction you want it to, it will not be pushed. One hundred kilograms of pork refusing to move will not be shoved. There will be times when you need to move pigs from one pen to another, load them up a ramp or into a trailer. Time for a little pig psychology.

It is always preferable to arrange circumstances where a pig will move willingly. Sometimes gentle stroking and encouragement is sufficient to get a pig to move where you want it to go. Alternatively, food can be used to entice pigs. This will be particularly effective if lighter feeds are given at previous meals. Pigs are wary of unfamiliar situations. Try to allow them access to loading ramps in the days prior to loading, even to the extent of feeding them in these locations.

Loading Ramps

Loading ramps are a useful way of getting pigs on and off vehicles. Commercial loading ramps for pigs are two and a half pigs wide. According to the experts, this accommodates the natural herding instincts of pigs and allows for two pigs to move forward together. Our pigs have obviously never heard of this theory.

Small-scale pig keepers may often want to move only one pig at a time. In our experience it is preferable for small scale producers to build loading ramps that are only one pig wide. Somewhere between 55 cm and 75 cm will usually be sufficient, depending on your breed of pig. Make sure the ramp has solid walls so pigs are not distracted by things they see around them.

Commercial piggeries use a solid piece of timber or blocking board as a means of steering pigs in the direction they want them to go. This is based on the thinking that a pig will only move in the direction that it can see. We have found this method largely ineffective and have at times found ourselves and the piece of board flung in all directions as a pig tossed us aside in disgust.

Loading pigs onto our utility used to be a frustrating and exhausting exercise until we discovered the garbage bin method. Pigs can be loaded single handed with less stress on the pig and the handler using a large plastic garbage bin or bucket. Simply slip the bucket over the head of the pig, being sure that it is large enough to cover the eyes. The pig will try to reverse out of the bucket and can be steered, rear end first, into a pen, up a ramp or into the back of a trailer.

Cleaning Pens

Pig pens can be cleaned out daily. This only involves removing solids and adding them to the compost heap. You will not have any smell if this is done and it takes only a few minutes.

Pigs like to sleep in bedding material such as sawdust or straw. Week-old bedding material can be recycled in the defecating area and replaced with fresh material. If absorbent material is used in this way, pens may only need cleaning once per week. Absorbent material will collect the urine, thereby capturing a valuable source of nitrogen that would otherwise be lost. Some smell is associated with this method, however, it is just the type of earthy, composting smell you would associate with any decomposing material. Many people buy in mulch for use on garden beds and around trees. This can be used in the pig pen first, recycled to the compost and then used on the garden.

Any manure is likely to attract flies and pig manure is no exception. We have found an Efekto Fly Trap to be most effective in minimising this problem. Pigs will smell if pens are not cleaned regularly or if the pigs are fed a diet consisting solely of over-ripe fruit.

Removing Tusks

The tusks of a boar appear as protrusions under the sides of the lip at around twelve months of age. They seem to grow fairly slowly, although this may vary from one breed to another. Some pig breeders prefer to remove the tusks of even the most gentle boar to prevent accidental damage to handlers or other pigs.

By the time your boar is four to five years old and has developed very large tusks, he may be getting too large to be mated with your sows. In such cases you will be thinking of replacing him with a younger boar and removal of the tusks will be unnecessary. However, if you have a good-natured boar with desirable breeding characteristics you may want to keep him for a much longer period. In such cases, the effort and expense of removing the tusks may be warranted.

Removal of the tusks is usually carried out using razor wire. It will require the assistance of a vet as the boar will need to be restrained and sedated during the procedure. This is also a good opportunity to check the general health of the animal.

15. Animal Health

Veterinary Assistance

It is useful to introduce yourself to the local vet and gauge their reaction when you mention that you intend keeping pigs. Most large-animal vets are able to handle and treat pigs, but many will admit that they have not touched a pig since university. We find our dog requires veterinary assistance more often than our pigs. Even if you do not require their services, it is useful to get to know your vet and be able to ask questions about any general problems relating to the health of your animals.

It is also useful to know what is roughly normal for a pig. The body temperature of a pig is 39 to 40 degrees Celsius. The pulse rate is around 70 to 80 beats per minute and the breathing rate is generally between 10 and 16 breaths per minute. These figures will vary between individuals and pigs of different ages.

If you handle and observe your pigs regularly, you will soon know when something is not quite right. The old story about a happy pig having a curly tail is also true.

Small-scale pig breeders are often interested in natural farming methods. Pat Coleby has spent her life researching the nutritional requirements of farm animals. She has written several texts on the subject (see bibliography). These make very interesting reading and contain information on using natural vitamins and minerals to help solve health and disease problems of a range of farm animals, including pigs.

External Parasites

Like most animals with hair or fur, pigs are subject to various external parasites. Once again, we have more problem with these on our dog than we do on our pigs. Pigs have a very thick, tough skin that is very difficult for ticks and lice to bite through. Heavy infestations can cause irritation to the animal and may cause it to scratch and graze the skin. Young piglets have soft skin and can be badly irritated by heavy infestations.

Pigs that continually rub themselves against fence posts or trees may have mange. Mange is caused by tiny mites. They can irritate pigs to such an extent that they rub the hair from their bodies and develop scabs around the head and neck. Poor nutrition and substandard housing can contribute to the incidence of mange.

A clean pig is a happy pig. Washing animals is a good way of ridding them of external parasites. Pay particular attention to the inside of the ears and the neck. Mites are likely to congregate in these areas. A pure soap and very mild disinfectant can be used, or heavy parasite infestations treated with pyrethrum-based products. Do not be tempted to treat your pigs with chemical flea and tick rinses, powders or ointments designed for cattle, dogs, cats or other animals. These may be based on organophosphate chemicals that can leave residues in pork and hams.

Any treatments for external parasites should be accompanied by a thorough cleaning and treatment of the animal's housing. Where bedding material is being used this should be replaced. Sprinkling bedding material with derris dust will also deter parasites.

Pigs gleam in the sun after a bath and love nothing better than a rub with vegetable oil. This is another useful way of treating parasites as they suffocate under the coating of oil. Watch that your pigs don't bask in the sun after you apply the oil or they will get sunburnt. If you provide a rubbing post covered with a well oiled hessian sack, the pigs will oil themselves. Do not be tempted to use old engine oil as suggested in some books. This may contain contaminants that will be transferred to your pigs via the skin. Ordinary cooking oil is acceptable. Pigs naturally shed hair during warmer weather and develop a thicker coat during the cooler months.

Other skin problems may be the result of nutritional deficiencies. Adequate grazing on fresh, green pasture and use of kelp supplements should prevent such occurrences.

Internal Parasites

Like most animals, pigs are subject to a variety of intestinal worms. These worms are not generally those found in cattle, sheep or dogs, but are those found in common with poultry. As poultry and pigs are commonly kept together this can be seen as an advantage or a disadvantage. Poultry can pass on worms to your pigs and vice versa, but equally any treatment preparations are suitable for both groups of animals.

Pigs infested with worms will be continually ravenous and fail to gain weight and grow as rapidly as they should. They will be less active and their coats will be dull. These are generally the first signs of infestation. It may be possible to see worms in the animal's droppings. By the time this is evident, the infestation is a severe one.

We have read and tried various treatments for worms, including adding vinegar or powdered sulphur to daily rations and feeding tansy to our pigs. In most instances the pigs would simply not eat what was provided or we were unconvinced that the treatments worked. Read and try every method you can to find something that works for you.

Pre-war texts often refer to the use of turpentine for worm control. The following details have been converted to current systems of measurement. Keep pigs without food for 12 hours, then give them a teaspoon of turpentine mixed in 150 mL of linseed oil. Follow up with a dose of castor oil within 24 hours.

Garlic and pumpkins (fed seeds and all) are said to be a natural vermifuge. You would need to grow your own to make their regular inclusion in your pigs' diet affordable. In *Farming Naturally and Organic Animal Care,* Pat Coleby reports successfully controlling worm infestations in goats and other animals by the addition of copper sulphate to the diet.

Most worms are transferred from the droppings of infected animals into the soil. The worm eggs are then taken in by animals feeding on infected pasture.

We have settled on the philosophy that prevention is better than cure. Our pigs and paddocks are largely worm free and maintained that way by the following routine. Any new pigs (or chickens) that are brought to the property are placed in separate pens and wormed with conventional worming preparations.

This isolation period usually lasts for about one week, until all the worms have been expelled. It also gives the animal time to settle into its new environment and allows it to be checked for external parasites. As the animals are confined during this time it is a simple matter to collect and remove all manure from the pens. The pen is then cleaned and treated with disinfectant.

Paddock grazing occurs on a rotation cycle of at least three months. This means that each paddock will actually be vacant and regenerating for at least six months before pigs are reintroduced. If you have room for more than three paddocks all the better. The cycle of worm infestation can be broken by long rotation periods.

Many people will disagree with the chemical treatment of pigs for internal parasites. However, an animal badly infested with worms is very distressed. We are currently trialling the use of insecticide-grade diatomaceous earth for intestinal worm control. This product is widely used by organic farmers in the United States for control of both internal and external parasites. Only a small quantity is required. Monthly doses of 10 to 20 grams, mixed with dry feed, are adequate. While diatomaceous earth is not toxic to mammals, care should be taken not to inhale the dust or get it in your eyes. Do not be tempted to use the diatomaceous earth used in swimming pool filters. This is not the same product and is quite dangerous.

By using these methods, we limit chemical treatment to introduced stock and rarely find the need to treat animals that will soon be used as meat for human consumption. If you are not introducing new animals to your property, an initial treatment combined with an effective paddock rotation cycle will minimise internal parasite problems.

Pig Diseases

Many animals, including pigs, can carry a range of serious diseases and parasites that can be transferred to humans. You may think that all your animals are completely healthy and disease free, but it is wise to follow normal hygiene practices when handling any animal. Always thoroughly wash your hands after feeding or handling your pigs. We have set up an old sink close to our animal pens. Liquid soap to which we have added a mild disinfectant is always available. Washing our hands after handling any animal is simply part of our normal routine. It is also good practice to disinfect animal pens regularly and to thoroughly clean food and water containers

Leptospirosis, parvovirus (not the same type that affects dogs), brucellosis and a number of other pig diseases can cause spontaneous abortions and the production of mummified or stillborn offspring or weak piglets that die soon after birth. Such serious disease problems are far beyond the scope of this book. Expert advice is required to diagnose such diseases, since the infected animal may show no outward signs of the disease.

Always isolate any suspect animal and call for veterinary assistance when disease problems are suspected. Small-scale pig breeders are less likely to have problems of this nature if they buy quality stock from a reputable source. Hygiene is also important. Rats and mice can transmit diseases. Store pig food in lidded containers to keep the rodent population down. Commercial intensive producers tend to vaccinate for disease problems.

The incidence of serious pig diseases varies from region to region. When transporting pigs long distances, particularly across state borders, it may be necessary to produce certification as to the disease-free status of animals. Always check with the appropriate government body.

Pigs can get tuberculosis (Tb) from infected cows. If you keep pigs and cows together, make sure that your herd is disease free. If you purchase pigs from a property that has cows, it may be prudent to ask questions.

Scouring or diarrhoea can occur in pigs of all ages as a result of *Escherichia coli*, coccidiosis or salmonellosis. A sudden change of diet (as occurs at weaning) and stress associated with transport, over-crowding or changes within pig hierarchies are also contributing factors.

Hygiene and good pig housing are major factors in reducing the incidence of scouring. Any affected pigs should be isolated and placed in a clean, quiet environment. Reduce the food intake and provide plenty of clean, fresh water. Piglets are especially prone to dehydration. Adding a little glucose to the water can also be beneficial. An old-fashioned natural remedy often recommended can be made from the water of boiled dill seeds and a little Epsom salts.

16. Composting Manures

Pig manure and bedding materials make wonderful compost. You will read in some books that pig manure should not be used on the garden for fear of transferring disease. Much of this information comes from England and Europe, where serious animal diseases occur.

We believe it is wise to take precautions when handling any animal manure. All animal owners should have regular tetanus boosters. Always wash your hands thoroughly after cleaning animal pens, turning compost or handling animals. Hot compost all manures to ensure any pathogens and parasites are killed.

The nutrient content of pig manure varies according to the diet of the animal producing it. The bedding material used and how the manure is stored and treated will also affect the compost. The nitrogen, phosphorous and potassium (N:P:K) ratio of pig manure from commercial piggeries is around 18:7.5:4.5. Pig manure also contains magnesium, zinc, copper and a range of other trace elements. Most of the nitrogen and potassium is found in the urine, while the phosphorous is found in the solids. Bedding material that absorbs the urine makes a rich compost.

Nutrient levels in pig manure are high when compared with other manures. Pig manure is a useful compost activator, helping to bring about the rapid decomposition of weeds, soft branches and other materials that may otherwise take a long time to break down. We use a hot composting method. This involves combining manure and bedding material with soft garden prunings and other organic materials at hand.

The combined material is thoroughly wetted and turned each week. Material on the outside of the heap should be moved towards the centre during turning. Covering the heap with thick plastic helps to contain any heat produced and protects the heap from becoming over-wet during heavy rain. The resultant heap gets so hot it steams. If you really want to be scientific you can test the heat being generated by the heap using an old roast thermometer. Temperatures above 55 degrees Celsius kill most harmful organisms and weed seeds. However, it is not unusual for the temperature in the centre of the heap to reach closer to 80 degrees Celsius during the first few weeks of composting. Beautiful friable compost should be ready within six weeks by using this method.

Glossary

Once fully composted, manures are best used straight away on the garden. If the compost is to be stored for any time it should be covered and slightly compacted. Much of the nutrient of the compost will be leached out if it is stored uncovered and subject to rain.

Ad Lib Feeding regimen where pigs are given continuous access to feed. Used to hasten the growth rate of young pigs.

Baconer Pigs 18-26+ weeks or between 70 and 90+ kg live weight, slaughtered for hams and bacon.

Boar Male pig of any age.

Creep Feed Technique of giving piglets especially formulated foods prior to weaning. Used to help them make the transition form the sow's milk to solid food and help maximise growth rates.

Extensive Production Commercial raising of pigs either completely outdoors or with considerable access to outdoor grazing.

Farrowing Process of giving birth to piglets.

Gestation Period from conception to birth usually, 114 -116 days for pigs.

Gilt A young female pig yet to be mated.

Intensive Production Commercial raising of pigs indoors.

Lactating Sow Sow feeding a litter of piglets.

Litter A group of offspring born to the same sow at the same time.

Oestrus Period during the reproductive cycle when a sow is able to be mated.

Overlaying Accidental crushing of piglets by sow.

Porker Pigs 12 - 18+ weeks old or between 30 and 70 kg live weight.

Scouring Diarrhoea generally caused by contaminated food, disease organisms, stress or in the case of piglets, by rapid change from liquid to solid food.

Sow Female pig who has had at least one litter.

Suckers Piglets from birth to weaning at 6-8 weeks.

Vermifuge Substance given to control internal parasites (worms).

Weaners Pig 6-12 weeks old or between 10 and 30 kg liveweight.

Weaning Removal of piglets from the sow and transfer to solid diet rather than milk.

Bibliography

Beynon N. *Pigs: A Guide To Management.* Wiltshire, UK: The Crowood Press, 1993.

Boatfield G. *Farm Livestock.* Ipswich, UK: Farming Press Books., 1994.

Coleby, P. *Farming Naturally and Organic Animal Care.* Euroa, Vic: Night Owl Publishers, 1991.

Coleby P. *Natural Goat Care.* Richmond, Vic: Max A Harrel, 1993.

Genders R. (1979) *Pig Keeping.* London: Foyles Handbooks, 1979.

Gardner J, Dunkin A, Lloyd L. *Pig Production In Australia.* Sydney: Butterworths, 1990.

Lipscomb A. *Breeding and Management of Live Stock: Cattle, Horses, Pigs.* Sydney: Whitcombe and Tombs, 1945.

McBarron EJ. *Poisonous Plants: Handbook for Farmers and Graziers.* Sydney: Inkata Press, 1991.

McGregor P. *Growing Pigs.* NSW: Australian Pig Corporation, 1993.

Mollison B. *Permaculture - A Designers' Manual.* Tyalgum, NSW: Tagari Publications, 1988.

Porter V. *Practical Rare Breeds.* UK: Pelham Books, 1987.

Thornton K. *Outdoor Pig Production.* Suffolk: Farming Press Limited, 1988.

Other Information

Suppliers

Green Harvest is a mail order company dedicated to stocking Australia's most comprehensive range of organic gardening supplies. They stock diatomaceous earth, the Efekto Fly Trap and its replacement baits. They also carry an extensive range of pasture improvement species that can be purchased as seed in small quantities. For a comprehensive catalogue send a self-addressed business size envelope with two stamps to Green Harvest, 52 Crystal Waters, MS 16 Maleny, QLD 4552. For further enquiries phone (074) 944-676, fax (074) 944 578.

Your local produce store should be able to supply cracked corn, barley and other grains, commercially prepared pig meal or pellets, dolomite, sulphur and other dietary additives. Prices vary considerably and discounts are often available for bulk purchases.

Associations/Contacts

Australian Pig Breeders Society (APBS)
PO Box 199
Kiama NSW 2533

Australian Rare Breeds Trust
Secretary: Clive Lloyd
RMB 1715
Wangaratta VIC 3678

Macaranga Tamworth Stud
G and A McFarlane
PO Box 235
Ferny Hills QLD 4055

Myola Tamworth Pig and English Leicester Sheep Stud
J and L Mulholland
PO Box 1014
Kenmore QLD 4069

Pig Research Council
Department of Primary Industries and Energy
Canberra ACT 2600

Pork Industry Council NZ
PO Box 4048
Wellington
NEW ZEALAND

Rare Breeds Survival Trust
National Agricultural Centre
Kenilworth, Warwickshire CV2LG
ENGLAND

Botanical Names of Lesser-known Plants Mentioned

Arrowroot	*Canna edulis*	Lucerne	*Medicago sativa*
Babaco	*Carica pentagona*	Madeira Vine	*Anredera cordifolia*
Buckwheat	*Fagopyrum esculentum*		
Carambola	*Averrhoa carambola*	Maku lotus	*Lotus uliginous*
Chicory	*Cichorium intybus*	Oats	*Avena sativa*
Choko	*Sechium edule*	Okra	*Hibiscus esculentus*
Cobblers Pegs	*Bidens pilosa*	Panama Berry	*Muntingia calabura*
Comfrey	*Symphytum officinale*	Pig Weed	*Portulaca oleracea*
Cowpea	*Vigna unguiculata*	Pigeon Pea	*Cajanus cajan*
Crown Vetch	*Coronilla varia*	Pinto's Peanut	*Arachis pintoi*
Egg Plant	*Solanum melongena esculentus*	Pomegranate	*Punica granatum*
Feijoa	*Feijoa sellowiana*	Rapeseed	*Brassica napus*
Glen Joint Vetch	*Aeschynomene americana*	Red Clover	*Trifolium pratense*
		Rosella	*Hibiscus sabdariffa*
Granadilla	*Passiflora quadrangularis*	Shaw Creeping Vigna	*Vigna parkeri*
Jaboticaba	*Myrciaria cauliflora*		
Japanese Millet	*Echinochloa utilis*	Strawberry Clover	*Trifolium fragiferum*
Jerusalem Artichoke	*Helianthus tuberosus*	Tagasaste	*Chamaecytisus palmensis*
Lab Lab	*Lablab purpureus*	Tamarillo	*Solanum betaceum*
Leucaena	*Leucaena leucocephala*	White Clover	*Trifolium repens*

Index

Abattoirs 13
Animal Husbandry 21
Associations 70

Background 7
Baconer 19, 57
Berkshire 16
Bibliography 69
Birth 45-47
Botanical Names 71
Breeders 70
Breeding 41
Breeds 16, 17
Breeds American 17
Breeds English 16
Breeds European 17
Butchering 13

Chickens 10
Cleaning 22, 23, 60
Commercial Production 22
Compost 10, 67
Conversion Rate 9
Costings 13, 19, 25
Councils 12
Cow 10

Development 57
Disease 15, 22, 65, 66
Duroc 17

Farrowing 45-47
Farrowing Rail 45
Feeder 33
Feeding 22, 24-34
Feet 12
Fencing 40
Food Sources 25, 26
Food Types 25-34
Food Values 31, 32
Foods 19, 24-34
Fruit 10, 27, 28

Gestation 9
Gestation Chart 42, 43
Gilts 19

Glossary 68
Gloucester Old Spot 16
Grazing 12
Growth & Development 57, 58
Growth Enhancers 14
Growth Measurements 58

Hampshire 17
Handling 59
Health 60
Hormones 14
Housing 37-39
Hybrid Pigs 15, 16

Introduction 8

Kunekune 17

Lactating Sows 32
Landrace 17
Large Black 16
Large White 16
Lifespan 22
Litters 9
Loading Ramps 60
Local Authorities 12

Management 59
Manure 10, 67
Mating 44, 45
Micro-pigs 17
Middle White 17
Milk 22

Nose Ring 23

Orchards 28
Organic Pork 13
Ovulation 9

Paddocks 12
Parasites 62, 65
Pedigree 15, 16
Pens 23, 37, 38
Piglets 47, 48
Plants 27, 28, 71

Poisons 30, 31
Porcine Somatotropin 14
Porker 19, 57
Production 9, 14, 19, 22, 41-48
Production Free Range 22
Production, Commercial 22
Purchasing 19, 20

Quality 19, 20

Rare Breeds Association 20, 70
Refrigeration 14
Registration Certificate 18
Reproduction Age 19
Reproduction Rate 9

Saddleback 16
Schools 21
Sexual Maturity 19
Shires 12
Soil Compaction 23
Starting Out 12
Storing Food 33
Sucker 57

Tail Clipping 22
Tamworth 16
Teeth Clipping 22
Temperature 62
Transfer Certificate 18
Transgenic Pigs 11
Tusk Removal 61

Vegetables 10, 26, 27
Vision 10

Water Catchment 35
Water Tanks 35
Watering 35
Weaner 19, 57
Weights 13, 57, 58
Worming 31, 32